Cool Careers in SPACE SCIENCES

D1529793

Sally Ride
Science

CONTENTS

Sally

Christopher

Sandra

INTRODUCTION 4

ASTROBIOLOGIST
Christopher Chyba 6

ASTRONAUT
Sally Ride 8

ASTRONOMER
Sandra Faber 10

ASTROPHYSICIST
Geoffrey Marcy 12

COMMUNICATIONS ENGINEER
Ann Devereaux 14

GEOLOGIST
Joy Crisp 16

Geoffrey

Ann

Joy

Carolyn

Neil

Pascal

IMAGING SCIENTIST
Carolyn Porco 18

PLANETARIUM DIRECTOR
Neil deGrasse Tyson 20

PLANETARY SCIENTIST
Pascal Lee 22

PLANT PHYSIOLOGIST
Ray Wheeler 24

SPACE REPORTER
Marcia Dunn 26

VOLCANOLOGIST
Rosaly Lopes 28

ABOUT ME 30

CAREERS 4 U! 32

GLOSSARY and INDEX 34

ANSWER KEY 36

Ray

Marcia

Rosaly

What Do You Want to Be?

Is space exploration one of your goals?

The good news is that there are many different paths leading there. The people who explore space come from many different backgrounds and include physicists, geologists, and electrical engineers; microbiologists, medical doctors, teachers, and more.

It's never too soon to think about what you want to be. You probably have lots of things that you like to do—maybe you like doing experiments or drawing pictures. Or maybe you like working with numbers or writing stories.

SALLY RIDE
First American Woman in Space

The women and men you're about to meet found their careers by doing what they love. As you read this book and do the activities, think about what you like doing. Then follow your interests, and see where they take you. You just might find your career, too.

Reach for the stars!

Sally K Ride

"Astrobiology is about understanding life on Earth and whether there might be life anywhere else."

CHRISTOPHER CHYBA
Princeton University

Anyone Out There?

Our planet is bursting with life. But are we alone? Chris Chyba's job is to answer that question. As an astrobiologist, he searches for evidence of microscopic life on other planets and moons in our solar system. He thinks one of the best places to look is Jupiter's icy moon, Europa. Beneath Europa's frozen surface is a churning, salty ocean. On Earth, wherever there's water, there's life. Could there be life in Europa's ocean? That's what Chris hopes to find out.

Silver-Screen Scientists

When Chris was a kid during the 1960s, "I watched movies about monsters or other threats. Scientists were depicted as heroes. They could save the day because they understood how things really worked," Chris says. He was inspired. Today Chris works with other scientists and leaders from around the world to help stop the spread of nuclear and biological weapons. It's not only in the movies that scientists work to make the world a better place.

Chris has traveled to remote places, such as Siberia, to search for microorganisms deep in the ice. If living things can survive in extreme conditions on Earth, then maybe life could exist in extreme conditions on other planets or moons.

An astrobiologist studies how life got started on Earth and the possibility of life elsewhere in our solar system. Chris looks for signs of life on Jupiter's moon, Europa. Other **astrobiologists**

* study microorganisms that live in extreme conditions on Earth.
* listen for radio signals from other planets.
* look for water and primitive life on other planets.
* study comets and asteroids for clues to Earth's chemistry.

Future Astrobiologist?

Chris, like many other scientists, is curious, excited about making discoveries, and likes to travel. Think about what you have in common with Chris. Then discuss with a partner why those characteristics would make you a good scientist.

Did U Know?

Microorganisms recently discovered in Antarctica can hibernate in the ice for up to a million years.

Many Moons

Jupiter has more than 60 moons! Some are tiny, but others are as big as planets. Research Jupiter's four largest moons—Callisto, Europa, Ganymede, and Io. Then, create a fact box for each moon.

1. Each of the following adjectives describes one moon. Match it with the correct moon and create a title for each fact box such as "Icy _____."

    * Icy
    * Volcanic
    * Huge
    * Cratered

2. In each fact box, write a short description of the moon. Include facts you find interesting, such as Ganymede is the largest moon in our solar system.

Microscopic and Tough

Extremophiles are microscopic organisms that love *extreme* conditions. Many have Greek names. Using their names as clues, think about places where you might find these extremophiles.

halophiles (Greek *hals* = salt)   thermophiles (Greek *thermos* = hot)
xerophiles (Greek *xeros* = dry)   barophiles (Greek *baros* = weight or pressure)

Check out your answers on page 36.

NASA Now Higher-ing!

Sally landed her astronaut job after responding to an ad NASA placed in the school newspaper at Stanford. "The moment I saw the ad, I knew that's what I wanted to do—to go on that adventure," Sally says.

SALLY RIDE

Sally Ride Science

Net Gain for Science

Ever since she was a little girl, Sally Ride has loved sports and science. She kept her eye on the ball—playing tennis, volleyball, and football in the street with the neighborhood kids. She also kept an eye on the stars and planets—using a small telescope her parents gave her. Even though Sally led her college tennis team, she chose not to continue competing in tennis. Instead, she studied physics at Stanford University.

Two Firsts

Sally launched into history in 1983. She became the first American woman to fly in space and NASA's youngest astronaut. "I had an unbelievable opportunity to do something very few other people have a chance to do," she says. Sally recently launched her own science company. Her company's mission is to encourage students' interest in science, and to inspire them to reach for the stars.

"Astronauts are on a timeless quest. Exploration is as central to our lives as breathing."

An astronaut travels into space to explore our world and beyond. Astronauts come from many different backgrounds, including aviation, engineering, science, and teaching. And they use their different skills in space. Sally controlled the Shuttle's robot arm. Other **astronauts**

✴ conduct experiments inside the space station.

✴ release satellites into space.

✴ study Earth's oceans, weather, and geology.

Sally flew aboard the Space Shuttle *Challenger*.

Ironic Breakfast

Whether in space or in school, eating right means getting enough vitamins and minerals. One important mineral is iron (Fe). Many foods contain iron naturally. But iron is also added to many foods— investigate for yourself.

• Place one cup of dry, cold cereal (with 100% of the daily requirement of iron) in a recloseable plastic bag.
• Crush the cereal.
• Drag a magnet back and forth across the bag.

Do you see black fuzz sticking to the magnet? That's the iron. Team up with a partner and investigate why your body needs iron. Then put together a list of iron-rich foods.

Build Your Own Team

Living and working in space is challenging. Astronauts rely on their training and teamwork. And they rely on each other's strengths, including their backgrounds in math and science.

Create a list of your strengths that you could use as an astronaut. Then work with other students to mix and match strengths and build a mission team.

Good Company

Science isn't just a subject—it can lead to many different careers. If you started your own science company, what sort of company would it be? What would it do?

Check out your answers on page 36.

SANDRA FABER
University of California Observatories

Galaxy Gal

As a young girl, Sandy Faber loved to watch the night sky with her father. But it wasn't until she was in college that she really took stargazing seriously. Today Sandy is one of the most respected astronomers in the world. Her observations of galaxies have helped to explain how they form.

Seeing Stars

The road to stardom wasn't easy. On the first night that she observed the sky in college, Sandy fell off a telescope platform and got a concussion!

Telescopic Traveler

Sandy has never left Earth, but she's "traveled" billions and billions of miles into space. She uses some of the most powerful telescopes around. Some of the telescopes she uses include the Hubble Space Telescope, the Lick Telescope in California, and the Keck Telescopes in Hawaii. These telescopes allow Sandy to "go" into deep space and see galaxies forming. Now, that's far out.

Sandy uses the Hubble Space Telescope (right) to study galaxies.

An astronomer uses telescopes, satellites, and similar instruments to observe other worlds and celestial objects beyond Earth. Sandy studies how galaxies, such as our own Milky Way, were formed. Other **astronomers**

* investigate the mystery of dark energy that fills much of the Universe.
* explore the birth of stars.
* search for Earth-like planets outside our solar system.

Hello from Earth

Imagine you could send a message into space that would reach an alien civilization in another galaxy. What would your message say about Earth and its inhabitants? What are the most important things you think another civilization should know? Work with a team to write a message. Then share it with the class.

Now think about what you would like to know about an alien civilization. With your team, write down at least five questions. Discuss your questions with the class, and how they would help you understand alien beings from far, far away.

Future Astronomer?

Like most astronomers, Sandy is curious and creative. She likes to solve puzzles, explore, and work on teams. What do you have in common with Sandy? Discuss with a partner which of your qualities would make you a good astronomer, and why.

"Finding a new planet brings a tingly feeling to the back of my neck. My mind goes off thinking about what that planet might be like."

GEOFFREY MARCY
University of California, Berkeley

Universal Inspiration

When Geoff Marcy was fourteen, his parents bought him a telescope. At night he'd take it outside and look at the stars and planets. "Witnessing the beauty of the Universe was better than going to the movies," he says. Even though the telescope was small, Geoff could see Saturn and its rings. He could trace the movement of Saturn's largest moon, Titan. "To actually see the clockwork of the Universe in action just blew my mind away," Geoff says.

Planet Hunter

Today Geoff has one of the coolest jobs imaginable. He hunts for exoplanets—planets outside our solar system. Since 1995, Geoff and his team have tracked down more than 160 new planets. "It's very exciting to find planets orbiting other stars. We compare them to the planets that orbit our Sun, especially Earth. We'll learn whether our solar system is common or rare in the Universe," Geoff says. Finding Earth-like planets would be a tantalizing discovery. Could there be life elsewhere in the Universe? That could really blow a lot of minds!

Geoff uses telescopes, such as this one, to measure light coming from stars. Changes in the properties of the light may mean that there's a planet orbiting the star.

An astrophysicist

investigates the Universe and how it works. Geoff searches for planets orbiting other stars. Other **astrophysicists**

* study how stars are born and die.
* use computers to simulate the origins of the Universe.
* observe faraway galaxies.
* search for other solar systems.

Sitting in front of an observatory, Geoff looks at a globe of Mars.

Good Morning, Planet W!

It takes 24 hours for Earth to spin around once on its axis. In that time, you'll see one sunrise. Things can be different on other planets, especially exoplanets—planets beyond our solar system. Imagine you're on vacation, visiting four different exoplanets. In a 24-hour period, how many sunrises would you see on each of the following planets?

Copy this chart and then fill in the blanks.

Planet	Number of hours to spin once on its axis	How many sunrises in 24 hours?
Earth	24	1
Planet W	12	?
Planet X	8	?
Planet Y	4	?
Planet Z	1	?

Your vacation would be filled with breakfasts!

About You

As a young boy, Geoff searched the night sky for constellations. Which constellations have you seen?

Is It 4 U?

Astrophysicists use powerful telescopes to

* study stars.
* figure out how the Universe works.
* discover new galaxies.
* try to understand black holes.

Choose one of these activities that you would like to do. Write a paragraph explaining why.

Check out your answers on page 36.

ANN DEVEREAUX
NASA Jet Propulsion Laboratory

Space Saver

Astronauts use laptops to communicate with folks back on Earth. But Ann Devereaux thinks space travelers need lighter, simpler computers-—ones they can wear. Using her skills as a communications engineer, she designed and built a personal computer that's worn as a headset. It has a mini-computer screen that comes down over one eye. It can display e-mails and documents, and even play videos.

Just Solve It

Ann is an engineer, but she isn't that crazy about math. "Whoever said you have to like math to be an engineer?" Ann says that what she likes about having an engineering background is that it gives her a "huge toolbox" of problem-solving techniques. That means Ann can "always do different kinds of work and not get bored," she says.

Ann designs simple, wearable computers. She has also built communications radios for the *Mars Reconnaissance Orbiter* (artist drawing at right).

An engineer uses math and science to design products, build machines, and solve problems. Ann improves space communications, but other **engineers**

* design and construct roads, buildings, bridges, and dams.
* test the quality and safety of products—from food to toys to rockets.
* create designs for new spacecraft and rovers.

Phone Fun

Signs of engineering smarts are all around you. Pay attention to the different kinds of phones you see—at home, at school, in stores, in offices. Make a list of them. What did engineers consider when they designed the phones? What features did they add to make them useful in different situations?

Now have some fun creating a Phone Features poster. Include the different phones and describe their purposes and features. Don't forget to sketch the phones and label their features. Do some engineering and come up with a "future phone." What does it look like? What are its new functions?

Hello at Light Speed

Scientists communicate across space using radio waves. Radio waves travel at the speed of light, about 300,000 kilometers (186,000 miles) per *second*!

It's a long way to other planets, such as Mars. Sending and receiving messages can be slow. When the *Phoenix* spacecraft arrived at Mars, the spacecraft was about 276,000,000 kilometers (about 171,000,000 miles) from Earth.

* How long did it take *Phoenix's* first message to get from Mars to Earth in seconds?
* How long did it take to get a message from Mars to Earth in minutes?
* How many minutes would it take to send a message from Mars to Earth and then back to Mars?

Use a calculator and do some out-of-this-world math.

About You

Ann likes to build computers. What do you like to build?

Check out your answers on page 36.

JOY CRISP

NASA Jet Propulsion Laboratory

Rocks Rock

What is it about rocks that Joy Crisp likes so much? It's uncovering their mysteries. "I really enjoy collecting rocks that were once liquid rock—magma. I try to figure out how they erupted and the conditions in which they formed," Joy says. So it's perfect that Joy's job enables her to investigate the mysteries of rocks on Mars. The rocks hold clues about Mars's past.

Red Rover

Joy is an expert on the geology of Mars. She was a good choice to be the lead scientist for the Mars Exploration Rover mission. This mission successfully landed two robotic rovers, *Spirit* and *Opportunity*, on opposite sides of Mars. The rovers' discoveries made headline news around the world. They found evidence of past water—lots of water—on Mars, including a shallow, salty sea. Could Mars have been wet enough, long enough for life to start? That's a question for future missions to answer. Meanwhile, you can bet that Joy will be happy analyzing all the data from her rovers.

Joy sits among the rocks she loves to investigate.

The Mars Exploration Rover *Opportunity* simulated Mars travel before the mission.

A geologist studies the history and composition of rocks and soil on Earth or other worlds. Joy studies rocks on Mars, but other **geologists**

* study fossils inside rocks to understand past life.
* look for water, minerals, or diamonds underground.
* work on environmental pollution and erosion problems.

Spirit **has snapped many photographs of Mars. But this one's not quite real. NASA added an image of** *Spirit* **to one of the rover's own photographs.**

Ha Ha

Q. What do you get when you put wheels on a rock?

A. Rock and roll.

Rock Groups

Geologists categorize rocks into three groups—igneous, sedimentary, and metamorphic. Look up how each rock group forms and how you can tell them apart. Then, based on what you've learned, make a fact card for each group.

Now, snoop around your school or neighborhood and look for rocks. Can you identify the different rock groups? Which type of rock is most common?

About You

Joy discovered her love for geology in college. But as a girl she loved English, reading, and math. What are your favorite subjects?

CAROLYN PORCO
Space Science Institute

Extra-terrific! When Carolyn was working on the next *Star Trek* movie, a model of ET was in the studio.

Carolyn takes a break from her mission of unraveling the mysteries of Saturn.

Ring Around the Planets

Call her a scientist with a thing for rings. Carolyn Porco studies the magnificent sheet of rings that surrounds Saturn—as well as those around Jupiter, Uranus, and Neptune. Her work has helped to explain what rings are—billions of bits of ice zipping around in the same direction as they orbit a planet. Her work has also helped to explain how rings interact with moons to create ringlets.

Hello, Saturn

Carolyn also keeps an eye on Saturn with her team. "We never tire of seeing this gorgeous giant," she says. The team chooses which images the *Cassini* spacecraft takes of Saturn and its rings and moons, including mysterious Titan. As *Cassini* orbits Saturn, it is giving scientists their closest views yet of the ringed planet.

An imaging scientist

studies pictures captured by spacecraft and Earth telescopes to learn more about objects in space. Carolyn studies planets with rings. Other **imaging scientists**

✳ examine gullies on Mars for evidence of flowing water.

✳ study volcanic plumes rising from Jupiter's moon Io.

✳ track swirling storms on Saturn.

✳ pick landing spots for future Moon missions.

Saturn is so huge that 750 Earths could fit inside it.

Wide, Wide World

Our planet is huge, right? Compare Earth to its neighbors in space. Earth's diameter is 12,756 kilometers (7,926 miles). Knowing this, calculate the diameters of Mars, Saturn, and our Sun—in kilometers, and in miles. If you want, you can use a calculator.

- The diameter of Mars is about one half that of Earth.
- The diameter of Saturn is about nine times that of Earth.
- The diameter of our Sun is about 109 times that of Earth.

Make a list of the three planets and our star, the Sun. Put them in order from smallest in diameter to largest in diameter. Where does our planet rank?

Think About

Carolyn was a consultant on the science fiction movie *Contact*. It was about searching for life beyond our planet. Do you think life exists on other planets or even moons? Why or why not?

Future Funnies

Q. Why do astronomers think Saturn is married?

A. Because it wears a ring.

Q. Why wouldn't you want to give Saturn a bath?

A. It would leave a ring around the tub.

Check out your answers on page 36.

"It's fun to think about the Universe and all the things we don't yet understand. I look at the unknown and say, I want to know it."

NEIL deGRASSE TYSON

American Museum of Natural History
Hayden Planetarium

Starved for Space

Even when he was a boy, Neil deGrasse Tyson knew that he wanted to be an astrophysicist—a scientist who studies how the Universe works. He was eager to learn everything he could about stars, planets, and galaxies. So, he took classes at the famous Hayden Planetarium in his hometown, New York City. "I'm grateful for the scientists and educators who taught me at the planetarium," Neil says. "They fed my hunger for the cosmos when I was a kid."

Giving Back

During college, Neil researched star formation and investigated stars in spiral galaxies. Early in his career, Neil got the chance to return to the Hayden Planetarium. This time he was a staff scientist—then its director! "It gives me a good feeling to know that when kids come to the planetarium, they learn something that they can take with them into the rest of their lives," he says.

Science Athlete

Neil wasn't pinned down to only math and science. He was captain of his high school wrestling team and was undefeated. He also wrestled in college. Now Neil is too busy doing research and introducing others to the cosmos to wrestle. But he still likes to work out and stay in shape.

In a lecture, Neil shows his excitement about the cosmos.

An astrophysicist uses physics to study our solar system and the rest of the Universe. Neil studies how stars and galaxies form, writes books, and creates programs to educate the public about the Universe. Other **astrophysicists**

* study black holes.
* explore how gravity works in the Universe.
* try to understand how the Universe began.
* observe and learn about giant exploding stars called supernovas.

Neil is on the set of *Origins*, a PBS/NOVA TV show, which he hosted.

Space to See

Have you ever been to a planetarium? Pick a partner and take turns describing what you liked best. Was it the guided tour of our galaxy? Was it exploring constellations? Or was it learning about how stars are born? If you've never been to a planetarium, discuss with your partner what you would like to see and why.

About You

Neil was inspired by teachers and scientists at the Hayden Planetarium. Who or what inspires you?

Why? How? Where?

Neil says he enjoys science because it allows him to ask questions and investigate. What would you like to investigate about the Universe? Form a team, and create five questions that you'd like to investigate. Make sure each question begins with either *why*, *how*, or *where*. Here are some words to include in your question.

- Planet
- Sun
- Universe
- Formation
- Light-year
- Star
- Moon
- Gravity
- Galaxy
- Mars

Then each team chooses their number-one question. Ask the whole class the question. Can your classmates help you answer it?

PASCAL LEE
Mars Institute/SETI Institute

Ready, Set, Go!

Pascal Lee has high hopes of sending you to Mars. That's right. Pascal and his team are preparing a future mission—the first human mission to Mars. It will be the biggest achievement ever in space exploration history. One thing on Pascal's to-do list is learning how to drive rovers on the Red Planet. They're not real rovers, but souped-up Humvees. And they're not on the *real* Mars, but on an Arctic island that looks like Mars. "It's like a gigantic Hollywood set—a real Mars park," Pascal says.

Life on Mars?

Astronauts will drive rovers long distances across the Red Planet to gather clues. What mystery will they try to solve? Whether Mars ever was—or still is—wet enough to support primitive life. Investigating that question and preparing for human exploration of Mars are just two of Pascal's passions. He has also explored Mars-like terrain at the other end of Earth—Antarctica—where Pascal searched for meteorites. How cool!

Mars Mimic

The island is a rocky and cold desert, with no roads or trees. Perfect—just like Mars! Pascal experiments with different techniques and technologies for exploring the real Mars. How to navigate the tough terrain, and other lessons Pascal learns, will help future astronauts be better Mars explorers.

Pascal tests a space suit in his giant outdoor lab— Devon Island in the Arctic.

A planetary scientist studies the planets, their moons and rings, and other objects in our solar system. Pascal studies the history of water on Mars— and whether life could have developed there. Other **planetary scientists**

✳ study the atmosphere of other planets.

✳ analyze the makeup of rocky asteroids and icy comets.

✳ search for pockets of liquid water below the surface of Jupiter's moons.

The MARS-1 Humvee rover is designed to safely hold crews of up to four researchers.

Roving Mars

Robotic rovers exploring Mars are operated remotely by scientists back on Earth. The rovers usually have a main body to hold a computer; a camera to snap pictures; science instruments to analyze the air, soil, and rocks; and solar panels to generate electricity to run on. Someday, astronauts will explore Mars. Imagine you're an engineer. Get to work and design a rover! Be sure to label its parts and include a caption that describes what it does. Here are some things to think about. It must

- go over rocks, into craters, and through deep sand.
- withstand below-freezing temperatures.
- support heavy weight.
- send data back to Earth.
- store food, water, oxygen, and other supplies.
- use renewable energy.

Building a Better Buggy

Pascal's Humvees are nothing like the open buggies that astronauts drove during the Apollo missions on the Moon. Instead, the Humvees are big, enclosed, and comfy—just like real Mars rovers will be. "It's really the difference between traveling in a golf cart and a camper," Pascal says.

If You Can Dream It . . .

When Pascal was a young boy in Hong Kong, he dreamed of space travel and exploration. He worked to make his dream come true. What are some things you can do today and the rest of the year to make your dream come true?

PLANT PHYSIOLOGIST

RAY WHEELER
NASA Kennedy Space Center

> "It's fun to imagine that the work you're doing today will help space colonists provide their own food and oxygen."

Plants in Space

Why would Ray Wheeler be studying plants at the Kennedy Space Center? What do plants have to do with space? Plenty! If you were on a mission to Mars, getting there, exploring, and returning to Earth would take nearly three years! After a while you would get pretty tired of freeze-dried soup, hamburgers, and mashed potatoes. Ray is working on how astronauts could one day grow their own plants as a source of fresh food in space.

Problem Solver

Ray has to solve many puzzles before astronauts can start planting space gardens. For instance, where will they get water? Plants take in water through their roots. They send it back into the air through their leaves as vapor. "So astronauts could collect the vapor, turn it back into liquid water, and pour it right back onto the roots." Recycling in space—clever.

Nature Lover

"When I was in the Boy Scouts, we would hike in the woods and observe trees and other plants," Ray says. He never imagined that exploring nature on Earth would lead him to exploring nature in outer space.

A plant physiologist

investigates the structures of plants, how plants work, and how they interact with their environment. Ray does experiments to see how plants will grow during space travel. Other **plant scientists**

* learn how plants use water.
* genetically engineer plants to grow in harsh climates.
* crossbreed plants to create new varieties.
* investigate how plants respond to light.

Ray checks on a type of lettuce that one day might grow in space.

H₂O to Go

Trees and other plants lose water to the air. This water becomes part of the water cycle. Do this experiment and see for yourself.

Cover the leaves of a branch with a clear plastic bag that has no holes. Tie the bag tightly with a string. After a couple of days, what do you see?

Now repeat the experiment with the same kind of plant, in a shadier or sunnier area. What do you notice this time? Is there more or less water vapor? Why? When you're done, be sure to remove the bag.

About You

Ray's interest in nature began when he was young. It led him to study plants in college. What do you like to learn about? Why?

Think About

Plant physiologists spend their time asking questions about plants and trying to find answers. What's your favorite tree or other plant? What would you like to know about it?

Adventure Seeker

Marcia liked science and math as a girl. But she especially liked words and writing. (She still does!) So, naturally, she studied journalism in college and then became a reporter. Before writing about space, Marcia wrote about medical news.

MARCIA DUNN
The Associated Press

Watching History

When NASA launches a spacecraft, Marcia Dunn has a front-row seat. Her office is only three miles from the Kennedy Space Center's launch pads. That's about as close as you can get. Marcia is a reporter. She writes news articles about space and NASA. She has reported on about 90 launches and interviewed space's biggest stars, from Neil Armstrong to Eileen Collins. "I love writing about the space program. No two days are alike," Marcia says.

Asking Questions, Getting Answers

Some topics Marcia writes about are very technical, such as NASA's plan to return astronauts to the Moon. To understand the science she needs to know, Marcia reads and asks lots of questions. "Science can be learned. You don't need to be an Einstein to be able to understand it and appreciate it."

At a NASA news conference, Marcia asks a question.

A science writer explains complicated scientific or technical information to others. Marcia writes about space for The Associated Press, a service that sends news to many newspapers, radio and TV networks, and Web sites. Other **science writers**

* write articles for science magazines and technical journals.

* write manuals that tell people how to operate computers, planes, and even weapons.

* write newsletters sent out by high-tech companies.

About You

Most scientists are happy to answer Marcia's questions because they want people to learn about what they do. To whom do you ask questions? How do you learn about things you don't know?

Future Reporter?

Reporters don't make things up. But a little make-believe is okay when you practice writing. Imagine that a space mission has discovered life on another planet in our solar system. Write a science article about the discovery. Be sure to include *who* discovered it (and *what* "it" is). Also explain *when*, *where*, and *how* the discovery was made, and *why* it's important.

In Your Life

With a partner, find and read a newspaper story about science, or read an online article from *Science News for Kids*. Then take turns asking each other these questions.

* What was the article about?
* In what ways was the article interesting?
* What do you wish the reporter wrote more about?
* What, if anything, was hard to understand?

VOLCANOLOGIST

ROSALY LOPES
NASA Jet Propulsion Laboratory

Sizzle in Space

What does a volcanologist study in outer space? Rosaly Lopes studies the most volcanically active place in our solar system—Jupiter's moon Io. The small moon has more than 150 active volcanoes. That's fewer than Earth's 600 volcanoes, but Rosaly's research shows that Io's volcanoes spew out lava far hotter than any lava on Earth. Hot stuff!

Crunch Time

The information on Io comes from thousands of images sent back to Earth from the *Galileo* spacecraft that orbited Jupiter for 14 years. Rosaly was sad to say goodbye to *Galileo* in September 2003. The spacecraft was sent plunging into Jupiter's atmosphere to burn up.

On-the-Job Excitement

Rosaly grew up in Brazil, but her heart was at NASA from an early age. She says she always wanted to work for the space agency, because space exploration is "never just a job. It's always exciting!"

Rosaly helped discover more than 20 active volcanoes that had never been seen before on Jupiter's moon Io.

Erupting volcanoes splotch the surface of Io in this photo taken by the *Galileo* spacecraft as it sped past.

A volcanologist studies the geology and composition of volcanoes to try to predict when they're going to erupt. Rosaly studies volcanoes on other planets and moons, but other **volcanologists**

* try to predict volcanic eruptions on Earth.
* study earthquakes that sometimes occur when there's volcanic activity.
* investigate the origin of lava, which starts out as magma, or liquid rock, deep underground.

The *Galileo* spacecraft was about as long as a school bus and traveled 4.6 billion kilometers (about 2.9 billion miles).

4 U 2 Investigate

How do volcanoes work? Why do volcanoes erupt? What does the material that a volcano spews out tell us about what lies deep below its surface?

Martian Mega-mountain

Io's volcanoes are nowhere near the tallest in our solar system. That record belongs to giant Olympus Mons on Mars—it stands about 27,000 meters (88,583 feet) tall! Compare Olympus Mons to three volcanoes on Earth. Calculate the heights of these volcanoes—in both meters, and in feet. You can use a calculator.

* Mount Etna in Sicily is about ⅛ the height of Olympus Mons.
* Mount Fuji in Japan is about ½ the height of Olympus Mons.
* Mount Rainier in Washington state is about ⅙ the height of Olympus Mons.

Check out your answers on page 36.

About Me

The more you know about yourself, the better you'll be able to plan your future. Start an **About Me Journal** so you can investigate your interests, and scout out your skills and strengths.

Record the date in your journal. Then copy each of the 15 statements below, and write down your responses. Revisit your journal a few times a year to find out how you've changed and grown.

1. *These are things I'd like to do someday.*
   Choose from this list, or create your own.

   - Become an astronaut
   - Become an expert on one planet
   - Design living quarters for Mars
   - Work on a team that researches new ways to explore space
   - Live on the Moon or Mars
   - Design robots
   - Learn about the Universe
   - Design ways to grow food on the Moon or Mars
   - Design new ways to communicate with astronauts
   - Build new space exploration vehicles
   - Search for signs of life on other planets

2. *These would be part of the perfect job.*
   Choose from this list, or create your own.

   - Being outdoors
   - Making things
   - Writing
   - Designing a project
   - Observing
   - Being indoors
   - Drawing
   - Investigating
   - Leading others
   - Communicating

3. *These are things that interest me.*
   Here are some of the interests that people in this book had when they were young. They might inspire some ideas for your journal.

   - Watching movies
   - Being a team player
   - Working with my hands
   - Wanting to be a scientist
   - Watching the night sky
   - Tinkering with cars
   - Playing musical instruments
   - Looking at stars through a telescope
   - Writing stories
   - Reading
   - Investigating plants
   - Doing math
   - Playing sports
   - Hiking

4. *These are my favorite subjects in school.*

5. *These are my favorite places to go on field trips.*

6. *These are things I like to investigate in my free time.*

7. *When I work on teams, I like to do this kind of work.*

8. *When I work alone, I like to do this kind of work.*

9. *These are my strengths—in and out of school.*

10. *These things are important to me—in and out of school.*

11. *These are three activities I like to do.*

12. *These are three activities I don't like to do.*

13. *These are three people I admire.*

14. *If I could invite a special guest to school for the day, this is who I'd choose, and why.*

15. *This is my dream career.*

Careers 4 U!

Which career is 4 U?
Space Sciences

What do you need to do to get there? Do some research and ask some questions. Then, take your ideas about your future—plus inspiration from scientists you've read about—and have a blast mapping out your goals.

On paper or poster board, map your plan. Draw three columns labeled **Middle School**, **High School**, and **College**. Then draw three rows labeled **Classes**, **Electives**, and **Other Activities**. Now, fill in your future.

Don't hold back—reach for the stars!

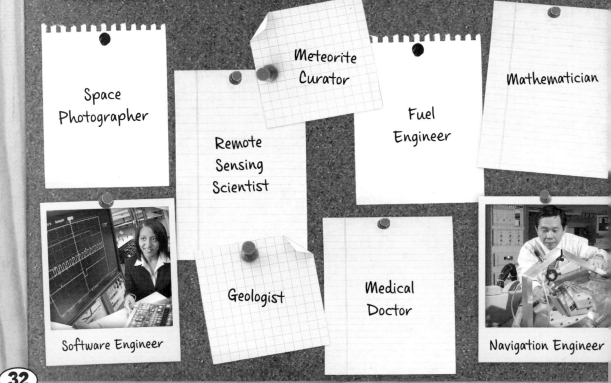

Space Photographer

Meteorite Curator

Mathematician

Remote Sensing Scientist

Fuel Engineer

Software Engineer

Geologist

Medical Doctor

Navigation Engineer

Planetary Scientist

Chemist

Microbiologist

Robotics Engineer

Physicist

Flight Engineer

Space Reporter

Astrobiologist

Inventor

Astronaut

Environmental Engineer

Rocket Scientist

Materials Engineer

Test Pilot

Biomedical Engineer

Electrical Engineer

Astronomer

asteroid (n.) A small rocky object that orbits the Sun. Thousands of asteroids orbit in a region called the Asteroid Belt, which lies between the orbits of Mars and Jupiter. However, some have been found in other orbits, including some that cross Earth's orbit. (pp. 7, 23)

astronomer (n.) A scientist who studies celestial bodies, including planets, stars, galaxies, and other astronomical objects. (pp. 10, 11, 19)

atmosphere (n.) A layer of gases surrounding a planet or moon, held in place by the force of gravity. (pp. 23, 28)

biological (adj.) Having to do with living things. Biology is the study of living things—plants, animals, and microscopic organisms—their behavior, physiology, development, and interactions. (p. 6)

chemistry (n.) The study of the elements and the ways in which they interact with each other. (p. 7)

comet (n.) A small body made up mainly of ice and dust, in an elliptical orbit around the Sun. As it comes close to the Sun, some of its material is vaporized to form a gaseous head and extended tail. (pp. 7, 23)

constellation (n.) A group of stars that appears to form a recognizable shape in the sky. (pp. 13, 21)

cosmos (n.) The Universe regarded as a complex and orderly system. (p. 20)

crater (n.) A bowl-shaped depression on the surface of a planet or moon caused by the impact of another body such as an asteroid or comet. (p. 23)

diameter (n.) The distance between two opposite points on a circle. It's twice the radius. (p. 19)

engineering (n.) The application of science and mathematics to design structures, such as bridges and wind turbines, and products, such as cell phones and biofuels. (pp. 9, 14, 15)

erosion (n.) The weathering down of rock by the action of wind or water. (p. 17)

galaxy (n.) A large collection of stars bound together by gravity. Our Sun is one of many stars in the Milky Way galaxy. (pp. 10, 11, 13, 20, 21)

journalism (n.) Writing designed for publication in a newspaper or magazine that is usually a direct presentation of facts or events without interpretation. (p. 26)

magma (n.) Molten rock beneath a planet's crust. (pp. 16, 29)

microorganism (n.) (also known as a microbe) A form of life, usually single-celled, that is too small to be seen without a microscope—including bacteria, some fungi, and some protists. (pp. 6, 7)

Milky Way (n.) Our own galaxy. It has a spiral shape, and our Sun is one of its billions of stars. (p. 11)

mineral (n.) A naturally occurring nonliving substance that has a characteristic chemical composition. Rocks are made up of mixtures of minerals. (pp. 9, 17)

observatory (n.) A building designed to hold telescopes, computers, and other equipment for astronomers to use for observing planets, stars, and other celestial objects. Most observatories are dome shaped and able to rotate so telescopes can look in any direction. (pp. 10, 13)

orbit (n.) The path of one body around another, as a result of the force of gravity between them. Examples are a planet's path around the Sun or a moon's path around a planet. (pp. 12, 18)

physics (n.) The branch of science that deals with matter, energy, and their interactions. Physics attempts to find laws, usually through math, which accurately describe a wide variety of phenomena throughout the Universe. (pp. 8, 21)

satellite (n.) An object that orbits another object in space. It also refers to something built to orbit Earth (for example, communications satellites and weather satellites). (pp. 9, 11)

Index

American Museum of Natural History, 20
Antarctica, 7, 22
Arctic, 22
Associated Press, 26, 27
asteroid, 7, 23
astrobiologist, 6, 7
astronaut, 8, 9, 14, 15, 22, 23, 24, 26
astronomer, 10, 11, 19
astrophysicist, 12, 13, 20, 21
atmosphere, 23, 28
aviation, 9

biological weapons, 6
black holes, 13, 21

Callisto, 7
Cassini, 18
chemistry, 7
comet, 7, 23
communications, 14, 15
communications engineer, 14
computers, 13, 14, 23, 27
constellation, 13, 21
Contact, 19
cosmos, 20
crater, 23

dark energy, 11
diameter (of Earth), 19

Earth, 7, 9, 10, 11, 12, 13, 14, 15, 17, 19, 23, 28, 29
earthquakes, 29
engineering, 9, 14, 15
environmental pollution, 17
erosion, 17
Europa, 6, 7
exoplanets, 12, 13
extreme conditions, 6, 7
extremophiles, 7

food in space, 24, 25
fossils, 17

galaxies, 10, 11, 13, 20, 21
Galileo, 28, 29

Ganymede, 7
geologist, 16, 17
geology, 9, 16, 17, 29
gravity, 21

Hayden Planetarium, 20, 21
hibernate, 7
Hubble Space Telescope, 10, 11

igneous rock, 17
imaging scientist, 18, 19
Io, 7, 19, 28, 29
iron, 9

journalism, 26
Jupiter, 6, 7, 18, 19, 23, 28

Keck Telescopes, 10

lava, 28, 29
Lick Telescope, 10
light-year, 21

magma, 16, 29
Mars, 13, 14, 15, 16, 17, 19, 21, 22, 23, 29
    Mars-1 Humvee Rover, 23
    Mars Exploration Rover, *Opportunity*, 16
    Mars Institute/SETI Institute, 22
    Mars Reconnaissance Orbiter, 14
    Mars, rocks on, 16, 17
metamorphic rock, 17
meteorites, 22
microorganisms, 6, 7
microscopic, 6
Milky Way, 11
minerals, 9, 17
moons, 6, 7, 18, 19, 23, 29
    Callisto, 7
    Europa, 6, 7
    Ganymede, 7
    Io, 7, 19, 28, 29
    Titan, 12, 18

NASA, 8, 17, 26, 28
NASA Jet Propulsion Laboratory, 14, 16, 28
NASA Kennedy Space Center, 24, 26
Neptune, 18
nuclear weapons, 6

observatory, 10, 13
Olympus Mons, 29
Opportunity, 16
orbit, 12, 18

Phoenix, 15
phone, 15
physics, 8, 21
planetarium, 20, 21
planetary scientist, 22, 23
planets, 6, 7, 8, 11, 12, 13, 15, 18, 19, 20, 21, 23, 29
    Jupiter, 6, 7, 18, 19, 23, 28
    Mars, 13, 14, 15, 16, 17, 19, 21, 22, 23, 29
    Neptune, 18
    Saturn, 12, 18, 19
    Uranus, 18
plant "water cycle" system, 24, 25
plant physiologist, 24, 25
plant physiology (system for getting energy), 24
plant structure, 24, 25
plants (in space), 24, 25

radio signals, 7
radio waves, 15
renewable energy, 23
rings, 18, 19, 23
robot arm, 9
robotic rovers, 16, 23
rock groups, 17
rocks, 16, 17, 23
    igneous rock, 17
    metamorphic rock, 17
    sedimentary rock, 17
rovers, 15, 17, 22, 23

Sally Ride Science, 8
satellite, 9, 11
Saturn, 12, 18, 19

Science News for Kids, 27
science writer, 27
sedimentary rock, 17
solar panels, 23
solar system, 6, 7, 11, 12, 13, 21, 23, 27, 28, 29
space reporter, 26
Space Science Institute, 18
Space Shuttle *Challenger*, 9
space station, 9
speed of light, 15
spiral galaxies, 20
Spirit, 16, 17
stars, 8, 12, 13, 20, 21
Sun, 12, 19, 21
supernovas, 21

telescopes, 8, 10, 11, 12, 13, 19
    Hubble Space Telescope, 10, 11
    Keck Telescopes, 10
    Lick Telescope, 10
temperature, 23
Titan, 12, 18

Universe, 11, 12, 13, 20, 21
University, 6, 8, 10, 12
    Princeton University, 6
    Stanford University, 8
    University of California, Berkeley, 12
    University of California Observatories, 10
Uranus, 18

vapor, 24, 25
volcanic plumes, 19
volcano, 28, 29
volcanologist, 28, 29

water, 23, 24, 25

CHECK OUT YOUR ANSWERS

ASTROBIOLOGIST, page 7

Microscopic and Tough
Halophiles live in highly salty environments.
Thermophiles thrive in very hot places.
Xerophiles exist in dry, arid locations.
Barophiles can withstand extreme pressure deep underwater.

Many Moons
- *Icy* Europa
- *Volcanic* Io
- *Huge* Ganymede
- *Cratered* Callisto

ASTRONAUT, page 9

Ironic Breakfast
Iron is essential to the human body. Most iron is found in hemoglobin—a protein molecule found in red blood cells. Hemoglobin carries oxygen to the cells where it's used as part of the energy-making process. Red blood cells can't function without iron. Too little iron leads to anemia—a condition that causes a person to feel weak or tired.

Iron-rich foods include whole grains; green, leafy vegetables; beans; nuts; eggs; lean meat; dried fruit; and shellfish.

ASTROPHYSICIST, page 13

Good Morning, Planet W!

Planet	Number of hours to spin once on its axis	How many sunrises in 24 hours?
Earth	24	1
Planet W	12	2
Planet X	8	3
Planet Y	4	6
Planet Z	1	24

COMMUNICATIONS ENGINEER, page 15

Hello at Light Speed
The first message from Mars to Earth took 920 seconds, or about 15 minutes.

- 920 seconds = 276,000,000 ~~kilometers~~ $\times \dfrac{\text{seconds}}{300,000 ~~\text{kilometers}~~}$

- 15.3 or about 15 minutes = 920 ~~seconds~~ $\times \dfrac{1 \text{ minute}}{60 ~~\text{seconds}~~}$

IMAGING SCIENTIST, page 19

Wide, Wide World
- Mars—12,756 kilometers × ½ = about 6,378 kilometers (about 3,963 miles)
- Saturn—12,756 kilometers × 9 = about 114,804 kilometers (about 71,336 miles)
- Sun—12,756 kilometers × 109 = about 1,390,404 kilometers (about 863,957 miles)

From smallest to largest
1. Mars
2. Earth
3. Saturn
4. Sun

VOLCANOLOGIST, page 29

Martian Mega-mountain
- Mount Etna—27,000 meters × ⅛ = about 3,375 meters (about 11,071 feet)
- Mount Fuji—27,000 meters × ½ = about 3,857 meters (about 12,656 feet)
- Mount Rainier—27,000 meters × ⅙ = about 4,500 meters (about 14,766 feet)

IMAGE CREDITS

NASA Johnson Space Center: Cover. © Rebecca Lawson Photography: p. 2 (Ride), p. 8 top. Courtesy R. R. Jones, Hubble Deep Field Team: p. 2 (Faber), p. 10. Courtesy Debra Fischer: p. 2 (Marcy), p. 12 top. NASA: p. 2 (Devereaux), p. 5, p. 8 bottom, p. 9, p. 11, p. 14, p.16 bottom, p. 19, pp 24-25, p. 28 bottom, p. 29, p. 33 bottom right. Courtesy Jared Pava for 92nd Street Y: p. 3 (deGrasse Tyson), p. 20 top. Maria Karras: p. 3 (Lopes), p. 28 top. Sally Ride Science: p. 4. NASA/JPL/Caltech: p. 17. Courtesy Delvinhair Productions: p. 20 bottom. Courtesy Diane Buxton for PBS/NOVA: p. 21. NASA Haughton-Mars Project: p. 22. Nick Lobeck: p. 24. Clara Lam: p. 30. Courtesy Northrop Grumman: p. 32 left. NASA Marshall Space Flight Center: p. 32 right. NASA/JPL/KSC: p. 33 top left. NASA /Tony Landis: p. 33 bottom left. © 2008 Georgia Institute of Technology/Photo by Rob Felt: p. 33 top right.